YOUR KNOWLEDGE HAS VALUE

- We will publish your bachelor's and master's thesis, essays and papers

- Your own eBook and book - sold worldwide in all relevant shops

- Earn money with each sale

Upload your text at www.GRIN.com and publish for free

Bibliographic information published by the German National Library:

The German National Library lists this publication in the National Bibliography; detailed bibliographic data are available on the Internet at http://dnb.dnb.de .

Imprint:

Copyright © 2018 GRIN Verlag
Print and binding: Books on Demand GmbH, Norderstedt Germany
ISBN: 9783668893559

This book at GRIN:

https://www.grin.com/document/456348

Alexander Mircescu

The Start of Life and Consciousness from Matter, Energy, Information, Space, Time, and Causality

Effects Based on Physics and Biology

GRIN Verlag

The Start of Life and Consciousness from Matter, Energy, Information, Space, Time, and Causality

Dr. Alexander Mircescu, Munich, Germany

1. The Four Levels of Physics and the Connection to Biology

In [1] we can see that physics takes place at four levels.

1. The first level is defined by the classical, Newtonian, physics. Here, matter, energy, space, and time are entirely separated from another. Furthermore, information and causality can be defined [1], also in a separated manner.

2. The second level is defined by the theory of special relativity and by quantum mechanics. According to special relativity space and time become coupled. According to quantum mechanics matter as defined by momentum and angular momentum becomes coupled to space; and energy becomes coupled to time. Hence, the second level leads to several couplings. Furthermore, information and causality can be also defined at the second level [1].

3. The third level is defined by the theory of general relativity and by quantum field theory. According to general relativity, spacetime becomes coupled to matterenergy. According to quantum field theory, matter and energy are generally defined by fields, such that matter and energy become entirely coupled. Hence, the third level leads to stronger couplings than the second level. According to [1], general relativity and quantum field theory can be described in a unified manner by causal nets, such that information and causality become also defined at the third level. At the third level it becomes clear, that the four fundamental forces strong force, electromagnetic force, weak force, and gravitational force are complemented by two further fundamental forces, namely by the quaternionic force and by the octonionic force [1].

4. The fourth level is defined by singularities in form of black holes defining local dynamic singularities, and in form of the big bang defining a global static singularity. Here, matter, energy, space, and time disappear as such, and are only represented by disconnected nets of infinite cardinality defining information and causality pieces [1]. Hence, the approach of information and causality applies also to the fourth level. At the fourth level, the couplings of matter, energy, space, and time are the strongest, since said four magnitudes disappear by a total coupling to information and causality which define the singularities alone.

We see that the higher physical level is, the stronger the couplings between matter, energy, information, space, time, and causality become. These couplings guarantee that the universe gets a stable structure and functionality by its elementary particles.

We also see that the first level which is physically the simplest level offers the biggest flexibility, since matter, energy, information, space, time, and causality are entirely separated from another, such that these magnitudes can be used without having to consider any restrictions according to couplings. This maximal flexibility becomes the starting point of biology which needs only the physics of the first level.

2. Matter, Energy, Information, Space, Time, and Causality at the Newtonian Level

In [1] we have seen how forces and torques lead to matter (as described by momentum and angular momentum), energy, and information, see also Figure 1 [1].

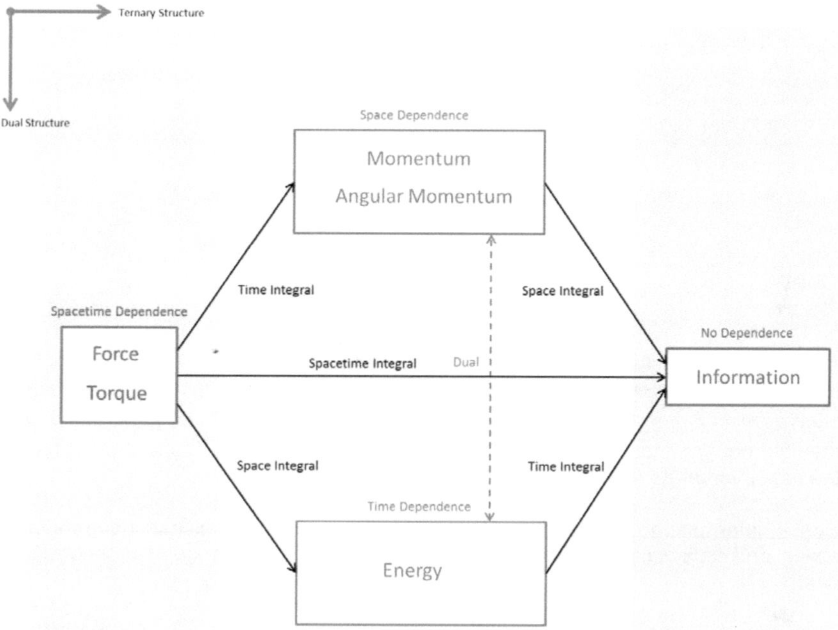

Figure 1: Relationship among force, torque, momentum, angular momentum, energy, and information

Starting from the force $\vec{F}(\vec{r}, t)$ and the torque $\vec{T}(\vec{\varphi}, t) = \vec{r} \times \vec{F}(\vec{r}, t)$ as functions of space $(\vec{r}, \vec{\varphi})$ and time t, the momentum $\vec{p}(\vec{r})$, angular momentum $\vec{L}(\vec{\varphi})$, energy $E(t)$, and information I are defined as [1]:

$$\int_{t_1}^{t_2} \vec{F}(\vec{r},t)dt = \vec{p}(\vec{r}),$$

$$\int_{t_1}^{t_2} \vec{T}(\vec{\varphi},t)dt = \vec{L}(\vec{\varphi}),$$

$$\int_{\vec{r}_1}^{\vec{r}_2} \vec{F}(\vec{r},t)d\vec{r} + \int_{\vec{\varphi}_1}^{\vec{\varphi}_2} \vec{T}(\vec{\varphi},t)d\vec{\varphi} = E(t) \ and$$

$$\iint_{t_1\vec{r}_1}^{t_2\vec{r}_2} \vec{F}(\vec{r},t)d\vec{r}dt + \iint_{t_1\vec{\varphi}_1}^{t_2\vec{\varphi}_2} \vec{T}(\vec{\varphi},t)d\vec{\varphi}dt =$$

$$\int_{\vec{r}_1}^{\vec{r}_2} \vec{p}(\vec{r})d\vec{r} + \int_{\vec{\varphi}_1}^{\vec{\varphi}_2} \vec{L}(\vec{\varphi})d\vec{\varphi} =$$

$$\int_{t_1}^{t_2} E(t)dt =$$

$$I$$

The information I is a scalar which is neither dependent on space nor on time, since the space dependence has disappeared by integrating the force $\vec{F}$ in the space interval $\Delta\vec{r} = \vec{r}_2 - \vec{r}_1$ and the torque $\vec{T}$ in the angle interval $\Delta\vec{\varphi} = \vec{\varphi}_2 - \vec{\varphi}_1$, and since the time dependence has disappeared by integrating the force $\vec{F}$ and the torque $\vec{T}$ in the time interval $\Delta t = t_2 - t_1$ [1].

Hence, information I is described by a scalar (by a number) which is independent of space and time, contrary to the force, torque, momentum, angular momentum, or energy [1].

I summarizes the momentum of the space interval $\Delta\vec{r} = \vec{r}_2 - \vec{r}_1$, the angular momentum of the angle interval $\Delta\vec{\varphi} = \vec{\varphi}_2 - \vec{\varphi}_1$, and the energy of the time interval $\Delta t = t_2 - t_1$ in an index I [1].

The information I is defined by the units [1]:

$$[I] = J\,s = V\,A\,s^2 = W\,s^2 = N\,m\,s = \frac{kg\,m^2}{s} := bit.$$

Information leads to causality which can be depicted in a causal net [1]. A possible causal net is shown in Figure 2 [1].

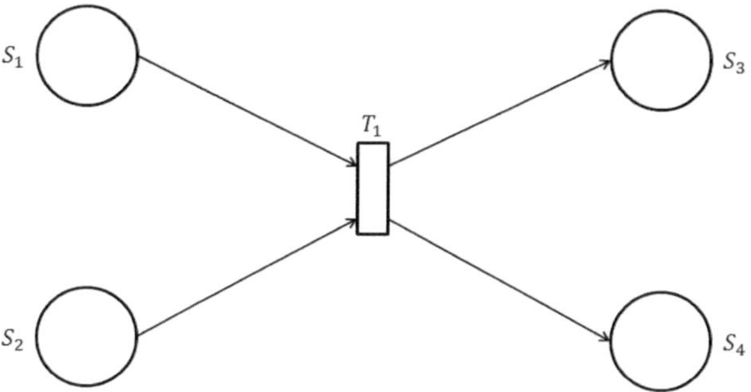

Figure 2: Example of a Causal Net

The states $S_1 - S_4$ represent the physical state of a system. We can identify said states with information as described above. Hence, the states $S_1 - S_4$ identify four information pieces $I_1 - I_4$.

The transition T_1 represents, together with the arrows, the causal relationship between the information pieces. The transitions define the interaction of the information pieces I_i. Hence, we have information pieces $I_{S1} \dots I_{Sn}$ before the interaction (S standing for "source"), such that said information pieces are connected by the transition to information pieces $I_{D1} \dots I_{Dm}$ after the interaction (D standing for "destination [1].

Otherwise stated, we have a vector of information pieces before the interaction [1]:

$I_S' = \begin{pmatrix} I_{S1} \\ \vdots \\ I_{Sn} \end{pmatrix}$, and we have a vector of information pieces after the interaction [1]:

$\vec{I_D} = \begin{pmatrix} I_{D1} \\ \vdots \\ I_{Dm} \end{pmatrix}$. The transition acts as a matrix [1] $\underline{K}$ projecting $\vec{I_S}$ to $\vec{I_D}$. The matrix $\underline{K}$ has

hence the structure $\underline{K} = \begin{pmatrix} k_{11} & \cdots & k_{1n} \\ \vdots & \ddots & \vdots \\ k_{m1} & \cdots & k_{mn} \end{pmatrix}$.

Hence, we have the equation [1]

$$\begin{pmatrix} I_{D1} \\ \vdots \\ I_{Dm} \end{pmatrix} = \begin{pmatrix} k_{11} & \cdots & k_{1n} \\ \vdots & \ddots & \vdots \\ k_{m1} & \cdots & k_{mn} \end{pmatrix} \cdot \begin{pmatrix} I_{S1} \\ \vdots \\ I_{Sn} \end{pmatrix}$$

showing how the information before the interaction is projected on the information after the interaction. Hence, the matrix $\underline{K}$ showing the interaction described by a transition is dimensionless. We identify the matrix $\underline{K}$ with causality, and conclude that causality is dimensionless, contrary to information [1].

The values momentum, angular momentum, energy, and information possess conservation laws [1]. We have

$$\vec{p}_1 = \vec{p}_2 \; (\textbf{\textit{Conservation of momentum}})$$

$\vec{p}_1$ describing the momentum before the interaction, and $\vec{p}_2$ after the interaction;

$$\vec{L}_1 = \vec{L}_2 \; (\textbf{\textit{Conservation of angular momentum}})$$

$\vec{L}_1$ describing the angular momentum before the interaction, and $\vec{L}_2$ after the interaction;

$$E_1 = E_2 \; (\textbf{\textit{Conservation of energy}})$$

E_1 describing the energy before the interaction, and E_2 after the interaction; and

$$I_1 = I_2 \; (\textbf{\textit{Conservation of information}})$$

I_1 describing the information of two particles before the interaction, and I_2 after the interaction.

Physics and chemistry (defined by the physics of the atomic shell) are founded on the values of forces and torques operating on matter, energy, and information, as observed in space, time, and causality. The structure and functionality of physics and chemistry can be described by systems and processes [3].

Definition 1 [3]: *System*: A system is a ten dimensional vector consisting of 3 dimensions of translational and rotational space, $\left((x, \varphi_{yz}), (y, \varphi_{zx}), (z, \varphi_{xy})\right)$, 3 complementary dimensions of space given by the overall momenta $\left((p_x, L_{yz}), (p_y, L_{zx}), (p_z, L_{xy})\right)$, 1 dimension of time (t), 1 complementary dimension of time given by the energy (E), 1 dimension of causality (k), and 1 complementary dimension of causality given by the information (I):

$$System = \begin{pmatrix} (x, \varphi_{yz}) \\ (y, \varphi_{zx}) \\ (z, \varphi_{xy}) \\ (p_x, L_{yz}) \\ (p_y, L_{zx}) \\ (p_z, L_{xy}) \\ t \\ E \\ k \\ I \end{pmatrix}$$

Definition 2 [3]: *Process*: A process is a nine dimensional entity consisting of a 3×3 matrix:

$$Process = \begin{pmatrix} Matter_{Processing} & Matter_{Transport} & Matter_{Storage} \\ Energy_{Processing} & Energy_{Transport} & Energy_{Storage} \\ Information_{Processing} & Information_{Transport} & Information_{Storage} \end{pmatrix}$$

such that each element of the matrix defines one dimension.

Now, the question arises how biological processes start from the physical/chemical processes leading to life and consciousness.

3. Life

In [2] a living (biological) system is founded on dissipative structures, as defined by irreversible thermodynamic processes. This approach is based on the work of Ilja Prigogine at the end of the 1960s. Ilja Prigogine observed that ordered structures can arise even in irreversible reactions far away from an equilibrium state. As an example, he points out that if we heat a pot filled with water and some plastic particles, then, contrary to thermodynamics, no total disorder of the plastic particles occurs. Instead, flows of the plastic particles are created, said flows defining a stable structure.

In order to define stable structures in dissipative systems, matter and energy must be constantly input and output [2]. This principle applies to any dissipative system, hence also to living systems.

Life is created far beyond any chemical equilibrium through a chain of instabilities which break down and are replaced by new instabilities as described by Darwin's evolution [2]. This evolution leads to higher and higher organisation degrees.

A living system is defined by the self-organisation of matter and by the evolution of biological macro molecules [2]. The evolution can be divided in three phases [2]:

1. The chemical phase where molecules are created.

2. The self-organisation of the molecules leading to simplest self-reproducing individuals.

3. The biological phase where the evolution of the species as defined by Darwin takes place.

The second phase defines the transition from dead to living systems.

We have seen that living systems demand that matter and energy are input and output. But living systems further demand the involvement of information and causality. Indeed, the self-organisation does not proceed in a random manner, but uses particular instructions, hence information. Hence, the randomness present in quantum mechanics is also present in mutation processes of the evolution, but said randomness is not isolated like in quantum mechanics; instead, the randomness is embedded in the selection process leading to causal chains.

Indeed, selection is carried out only in particular systems under particular defined restricting conditions. First, a solution is guessed; then, if the guess was successful, the information is stored; then, the stored information is used to proceed with guessing a better solution, and so forth [2].

In biology we have two main substances [2]: nucleic acids and proteins. The nucleic acid is transported to the place where the protein is synthesized. Hence, we have a process where a transport of nucleic acid takes place, where a processing between nucleic acid and protein takes place, and where the result of the processing is stored.

In [2] it is shown that only matter which is able to grow and which possesses auto catalytic properties allows the storage of information. (*Auto catalytic means that the chemical reaction is accelerated by a substance which is produced by the chemical reaction itself. Nucleic acids and proteins are examples of auto catalytic substances.*)

Matter can only perform selection processes when given external conditions are satisfied. The continuously growing and self-improving matter must never reach an equilibrium state, defined by the state which can be maintained with the minimal extent of energy; such a state is always aspired to by every chemical system. This is the clear distinction between biological and chemical systems [2].

In order to keep a system away from an equilibrium state, free energy must be continuously input to said system, such that work can be carried out by the system. This work defines the biological evolution such that certain types of matter are selected according to certain conditions, leading to the conservation of information carriers or entire systems, and to the death of remaining others [2].

The discussion in [2] shows that the nucleic acids and proteins together possess the properties for setting up living systems, such that the systems are able to reproduce themselves by cyclic processes, and such that harmful alternatives can be quickly disabled.

We can summarize our discussion from an abstract point of view as follows. Matter is continuously processed by cells which reproduce themselves. There are no distinct processes adapted to store matter for long times in special matter storage systems which can be accessed by different cells. Furthermore, matter is not transported in space by propagating through cells from outside like energy. Hence, only the value $Matter_{Processing}$ of Definition 2 is not zero.

Energy is continuously transported from outside to the cells. This energy is not stored for long times in special energy storage systems which can be accessed by different cells; instead, said energy transported to a cell is producing work in this cell. Furthermore, there are no distinct processes adapted to transform said energy among different types such that said types are processed in different manners; instead of processing the energy, the energy is directly performing work. Hence, only the value $Energy_{Transport}$ of Definition 2 is not zero.

Information is stored in the cell, enabling the cell to reproduce itself. This information is not processed by causally connecting different information pieces from large sets of different cells, and by always using newly constructed information pieces. Furthermore, information is not transported in space by propagating through cells from outside like energy. Hence, only the value $Information_{Storage}$ of Definition 2 is not zero. Hence, for living systems the process described by Definition 2 is more specific.

Definition 3: Biological *Process*: A biological process is a nine dimensional entity consisting of a 3 × 3 matrix:

$$Biological - Process = \begin{pmatrix} m_k & 0 & 0 \\ 0 & e_r & 0 \\ 0 & 0 & i_t \end{pmatrix}$$

which possesses only diagonal elements: m_k showing the causal processing of matter, e_r showing the spatial transport of energy, and i_t showing the storage of information in time.

In general, matter is coupled to space (see the definition of momentum and angular momentum), energy is coupled to time (see the definition of energy), and information is coupled to causality (see the definition of information and causal nets).

In biological systems we have second couplings: matter is further coupled to causality by m_k; energy is further coupled to space by e_r; and information is further coupled to time by i_t. Table 1 summarizes all couplings.

First Coupling	Substance	Second Coupling
Space	Matter	Causality
Time	Energy	Space
Causality	Information	Time

Table 1: Couplings of Matter, Energy, and Information with Space, Time, and Causality in Biological Systems

Hence, in biological systems, we have a causal connection of spatial structures, a spatial transport of time dependent processes, and a storage in time of causal structures. The presence of first and second couplings distinguish the biological systems from physical/chemical systems where only the first coupling is present. Therefore, biological systems are more complex than chemical processes.

Since living systems are defined by irreversible thermodynamic processes, when a living, hence non-equilibrium, system is destroyed, there is no possibility to reconstruct the destroyed system. Hence, if a transition from life to death takes place, there is no possibility to inverse the transition from death to life. It is, however, possible, to construct a new living system from the dead chemical substances.

4. Consciousness

Living systems perform evolution processes leading to constantly improving new living systems, as already depicted by Darwin in the evolution of the species. The action of a new living system, such that said living system is creating new causal relationships between the information pieces of itself and other living systems until death shall be defined as consciousness. Since biological processes are based on matter (as defined by momenta and angular momenta), energy, and information, the conservation laws for momenta, angular momenta, energy, and information also apply for biological systems.

The law of conservation of information implies that no information gets lost by any interactions [1]. Hence, the evolution processes conserve information when continuously producing new causal relationships until the living system dies. Therefore, the consciousness gathered by said causal relationships starts with the birth of the living system and ends with the death of the living system. Hence, consciousness as defined by the causal relationships constantly grows from birth to death. Since information is conserved, the death of a living system stops the growth of consciousness but conserves the state of consciousness present at the point in time when death occurs.

Since consciousness is coupled to interacting processes leading to causal relationships, consciousness can only be set up when a living system interacts with other living systems. Hence, when a living system is isolated, meaning that the sensors and actuators of the living system are blocked, then life continues but the state of consciousness is frozen and cannot grow any more. Only after the isolation of the living system is overcome, the growth of consciousness can continue. Hence, during a comatose state, a living system possesses life but the consciousness is blocked and frozen at the consciousness state corresponding to the time when entering the comatose state. When the comatose state is overcome, consciousness can continue; if the living system dies during the comatose state, then consciousness remains at the frozen level.

Bibliography

[1]: A. Mircescu: General Index Theory: Its Mathematical and Physical Structures, GRIN: Catalog Number: v423592, ISBN: 9783668691797, 2018.

[2]: M. Eigen: Stufen zum Leben, Serie Piper, 2000.

[3]: A. Mircescu: Systems and Processes Defined by the Substances Matter, Energy, and Information in the Existence Forms Space, Time, and Causality; GRIN: Catalog Number V366501, ISBN: 9783668455405, 2017.

YOUR KNOWLEDGE HAS VALUE

- We will publish your bachelor's and
  master's thesis, essays and papers

- Your own eBook and book -
  sold worldwide in all relevant shops

- Earn money with each sale

Upload your text at www.GRIN.com
and publish for free